Lost Scientific Literature of Bharat

By

Dr. Ravi Prakash Arya

AMAZON BOOKS, USA

in association with

INDIAN FOUNDATION FOR VEDIC SCIENCE
1051, Sector-1, Rohtak-124001, Haryana, India
Ph. Nos.: 09313033917; 09650183260
Emails: vedicscience@rediffmail.com
vedicscience@hotmail.com
Website: www.vedicscience.net

First Edition

Kali era: 5019 (c. 2017)

Kalpa era: 1,97,29,49,119

Brahma era: 15,50,21,97,9,49,119

ISBN No. 81-87710-89-6

Bharat, the cradle of the world civilisations on the earth, is the place of origin of Vedas and through them the place of origin of all other sciences of the world. It is proved by the fact that from Vedic period till the time of *Mahābhārata* war, science and technology were at its pinnacle. The seers of this land discovered all the laws underlying cosmic and physical creation and thereby developed a vast study material (literature) covering both the aspects of science, i.e. spiritual (metaphysical) and material (astrophysical & physical) sciences.

Who affords to forget those dark events of the human history and civilisation, when libraries at Nalanda, Vikramshila and Ujjain were set ablaze? Alexandrian library in the west too met the same fate at the hands of the enemies of civilisation. Had all the literature been preserved today intact, there would have been a different world in terms of history and sciences. In spite of this mass scale destruction of the books in the libraries, a huge number of Sanskrit manuscripts are still preserved in various museums and libraries of India. What to say of India, the libraries and museums of America and Europe are also holding around 1.5 million Sanskrit Manuscripts. The need is to collect, edit and publish this vast literature.

One more gruelling thing is that there is a misconception among learned and laity that Sanskrit is a language of rituals. Its literature generally deals with spirituality, philosophy, ethics, morality and religion. In the 19th century, when Swami Dayananda Sarasvati pronounced that Vedas are the storehouses of all true sciences, nobody, even the Sanskrit scholars in India was ready to believe this statement. Till 20 years back almost 99% of Sanskrit scholars were not able to tolerate that Vedas or Sanskrit contains positive sciences. Until date whatever scientific work is done on Vedas or in Sanskrit language, scholars from the scientific streams have taken the lead. Most of the

Sanskrit scholars in India due to their poor background of science, do not dare to take up any scientific studies. Moreover, the Sanskrit scholars in the West do not want to talk about science in Sanskrit out of prejudices and preconceived notions. They have a prejudiced thinking that the science is the prerogative of the modern west and the ancient language like Sanskrit have nothing to do with science. They associate past with darkness and primitive races. At the most, they take Vedic literature granted for religious or sacred literature. They forget that Sanskrit literature has never propagated religion at any stage or at any time and place. Sanskrit literature talks about Dharma and that Dharma is not different from today's science.

Ancient India is the cradle of different sciences and technologies. Just to find a small inkling of this fact, I quote here some citations from Hindu Superiority (2007:349-360). It is an undisputed fact that Astronomical, mathematical, medical, yogic and the military sciences have been very popular until present day. In Ancient India many other equally important sciences flourished which is evident from the remains of some of the most important achievements of the Indians. Mr. Elphinstone (p.133) says, "In science, we find the Hindus as acute and diligent as ever."[1]

Medical science in a flourishing condition presupposes the existence in an advanced state of several other sciences, such as. Botany, Chemistry, Electricity, etc. The Astra Vidyā (Military Science) presupposes the existence of the sciences of chemistry, dynamics, meteorology, geology, physics, and other cognate sciences in a much more advanced state than what we find them in at the present day; while the Vimāna Vidyā presupposes an intimate acquaintance with an equally great number of such sciences. The huge buildings of ancient India and "those gigantic

[1] Quoted by by Diwan Bahadur Harbilas Sarda (2007: 349)

temples hewn out of lofty rocks with the most incredible labour at Elephanta, Elora and at many other places," which have not only excited admiration but have been a standing puzzle to some people, could not have come into existence if the ancient Indians had not been masters of the science of engineering. The engineering skill of the ancients was truly marvellous. With all its advanced civilisation, modern Europe has yet to produce engineers able to build the Pyramids or to turn huge rocks into temples. Mons de Lesseps was no doubt an admirable representative of triumphant engineering skill and was an honour to France, but the only followed in "the footsteps of his predecessors,' who were equally great, and who, too, had at one time connected the Red Sea with the Mediterranean. Mr. Swayne (1887: 41) says, "A French engineer repeats the feat of the old native kings and the Greek Ptolemies in marrying by a canal the Red Sea to the Mediterranean, an achievement which will make the name of Lesseps immortal if the canal can only be kept clear of sand." The sands still maintain a threatening aspect.

As regards the Pyramids, the early fathers of the Church (Christian teachers before 500 AD.), believed them to have fallen from Heaven, while others in Europe believed them to have sprung out of the earth or to have been built by Satan and his devils.

The *Mahābhārata* shows that the ancient Indians had achieved wonderful advancement in mechanics. In the description of the Māyāsabhā (Exhibition), which was presented by Mayāsura to the Pāṇḍavas, mention is made of microscopes, telescopes, clocks, etc.

One would wonder to know the mechanism of the Māyāsabhā, which accommodated thousands of men, that it required only ten men to turn and take it in whatever direction they liked. There was the steam or the fire-engine called the Agni ratha.

There was powerful telescope like technology in ancient

India. One is mentioned in the *Mahābhārata*. It was given by Vyāsajī to Sanjaya at Indraprastha, in order to witness the battle going on at Kurukshetra.[2]

As regards the science of botany, Professor Wilson says, "They (the Indians) were very careful observers both of the internal and external properties of plants, and furnish copious lists of the vegetable world, with sensible notices of their uses, and names significant of their peculiarities." If the Akhbar-ul-Sadeeq (1887:7)[3] is to be trusted, a Sanskrit dictionary of botany in three Volumes was discovered in Kashmir in 1887.

Professor H.H. Wilson[4] says that artificial magnets and properties of loadstone were known to the Hindus. In the play *Mālatī Mādhava*, it is stated that the damsel drew Madhava's heart "like a rod of the iron-stone gem," which clearly shows that the Hindus were acquainted with artificial magnets as well as with the properties of the loadstone.

Professor H.H. Wilson further says, "The Hindus early adopted the doctrine that there is no vacuum in nature, but observing that air was excluded under various circumstances from space, they devised, in order to account for the separation of particles, a subtle element, or ether, by which all interstices, the most minute and inaccessible, were pervaded, a notion which modern Philosophy intimates some tendency to adopt, as regards the planetary movements, and it was to this subtle element that they ascribed the property of conveying sound: in which they were so far right that in vacuum there can be no sound. Air again is said to be possessed of the faculty of touch, that it is the medium through which the contact of bodies is effected either keep

---

[2] See *MahÈbhÈrata, BhÏ-ma Parva*, 2.10

[3] Quoted by Diwan Bahadur Harbilas Sarda (2007: 350)

[4] Quoted by Mrs. Manning, 1869. *Ancient and Mediaeval India*, W.H. Allen & Co. London . Vol. II, p. 209

them apart or impels them together. Fire, or rather light, has the property of figure – Mr. Colebrooke renders it of colour. In either case, the theory is true; for neither colour nor form is discernible except through the medium of light. Water has the property of taste, an affirmation perfectly true, for nothing is sensible to the palate until it is dissolved by the natural fluids."[5] This shows that the Indians were in no way behind the scientists of the nineteenth century.

The influence of the moon in causing tides was known to the Indians from the earliest times. *Raghuvañśa* (5.61) says:

तं तस्थिवांसं नगरोपकंठे।
तदागमारूढ़ गुरूत्व हर्षः।
प्रत्युज्जगाम कथकैशिकेन्द्रः।
चन्द्रं प्रवृद्धोर्मिरिवोर्मिमाली।।

That the Hindus were excellent observers and became great Naturalists becomes clear from Professor Wilson's note on a verse of the drama of *Mṛcchhkatika*, Chārudatta says:

> "The elephants' broad front, when thick congealed The dried-up-dew, they visit me no more."

Wilson says, "At certain periods a thick dew exhales from the elephant's temples. This peculiarity, though known to Strabo, seems to have escaped Naturalists till lately, when it was noticed by Cuvier (1835: 22 footnote)''.

Facts regarding diamonds, pearls, sapphires, etc., are mentioned with care, which show that the ancient Indians were thoroughly well-versed in the sciences and the arts relating to the fishery and to mining, and- the processes of separating and extracting various substances from the earth.

That the ancient Indians were masters of the sciences of chemistry, mechanics, meteorology is proved by one of the most wonderful of human achievements: the Vimāna Vidyā. The balloons of the Western world give us an idea of what Vimānas may have been like. 150 years ago a Vimāna was

---

[5] Quoted by Diwan Bahadur Harbilas Sarda (2007: 351)

considered an impossibility. But happily those days of Western scepticism are over, and a Vimāna; for its practical advantages, is looked upon as an ideal of scientific achievement. A European critic says, "Vimāna Vidyā was a complete science amongst the ancient Hindus. They were its masters and used it for all practical purposes."[6]

Vṛhad Vimanaśāstra by Maharishi Bharadwāja is a living example of this advance science in india.

This indicates their mastery of all the arts and sciences on which the Vimāna Vidyā is based, including a knowledge of the different strata and the currents of the atmospheric air, the temperature and density of each, and various other minor particulars. Vimāna Vidyā thus clearly mentioned in the Vedas. The *Yajurveda* (VI, 21) says:

समुद्रं गच्छ स्वाहा अन्तरिक्षं गच्छ स्वाहा देवं सवितारं गच्छ स्वाहा।।

Manu also says:

संशोध्य त्रिविधं मार्गं षडविधं च बलं स्वकम्।

सांपरायिककल्पेन यायादरिपुं शनैः।।

This science is said by some to have been a part of the more comprehensive science called "the Vāyu Vidyā" mentioned in the *Śatapatha Brāhmaṇa* (Kāṇḍa 11 and 14).

Prof. Weber (1878: 265) says, "*Sarpa Vidyā* (serpent science) is mentioned in the *Śatapatha Brāhmaṇa* (Kāṇḍa XIII) as a separate science and Viṣa Vidyā (science of poisons) in the *Āśvalāyana Sūtra.*" "Śivadāsa, in his Commentary of Chakrapāṇī, quotes Patañjali as an authority on *Lohasāstra*, or 'the Science of metals or metallurgy'."[7]

The Greeks derived their knowledge of electricity from

---

[6] Diwan Bahadur Harbilas Sarda, *Hindu Superiority*, Ed. Dr. Ravi Prakash Arya, International Vedic Vision, New York & Indian Foundation for Vedic Science, India, 2007.

[7] Quoted by Diwan Bahadur Harbilas Sarda (2007: 353)

India. Thales, one of the Greek sages, learned during his tour in India that when amber was rubbed with silk it acquired the property of attracting light bodies.[8]

Not only were the sciences of electricity and magnetism extensively cultivated by the ancient Indians, but they received their highest development in ancient India. The Vedantist says that lightning comes from rain. This can be easily demonstrated by the well-known experiments of Touilet and others: all these prove that Vedic sages perfectly understood all the electrical magnetic phenomena. The most significant proof of the high development of these sciences is to be found in the fact that they were made to contribute so much to the every-day comfort and convenience of the whole community, and that their teachings were embodied in the daily practices of the ancient Indians, which does the highest credit to their practical wisdom and their scientific temperament.

Sleep is necessary not only to enjoy sound health but to keep the body and soul together. The question now is in what way to sleep to derive the greatest benefit from this necessary operation of nature. Its solution by the ancient Indians not only proves them to have been masters of the sciences of magnetism and electricity but shows the spirit of Vedic tradition, which cannot be commended too highly for its readiness at all times and in all directions to adopt and assimilate the teachings of science. Every Indian is instructed by his or her mother and grandmother to lie down to sleep with the head either eastward or southward.

This practice is enjoined by the *Śāstras*. The *Ānhika, Tattva*, a part of our *Smṛti Śāstras*, says, "1. The most renowned Garga Ṛṣi says that man should lie down with his head placed eastward in his own house, but if he longs for longevity he should lie down with his head placed southward. In foreign places he may lie down with his head

---

[8] Diwan Bahadur Harbilas Sarda (2007: 353)

placed even westward, but never and nowhere should he lie down with his head placed northward."

2. Mārkaṇḍeya, one of the much revered Hindu sages says that man becomes learned by lying down with his head placed eastward, acquires strength and longevity by lying down with his head placed southward, and brings upon himself disease and death by lying down with his head placed northward."

There is a *śloka* in the *Viṣṇu Purāṇa*, which says, "Oh king! It is beneficial to lie down with the head placed eastward or southward. The man always lies down with his head placed in contrary directions becomes diseased."

In view of the above statements, it is not very difficult to conceive that the body of the earth on which we live is being always magnetized by a current of thermal electricity produced by the sun. The earth being a round body, when its eastern part is heated by 'the sun its western part remains cold. In consequence, a current of thermal electricity generated by the sun travels over the surface of the earth from east to west. By this current of thermal electricity the earth becomes magnetised, and its geographical north pole being on the right-hand side of the direction of the current, is made the magnetic north pole, and its geographical south pole being on the left-hand side of the same current, is made, the magnetic south pole. That the earth is a great magnet requires no proof more evident than that by the attractive and repulsive powers of its poles, the compass needle, in whatever position it is placed, is invariably turned so as to point out the north and the south by its two ends or poles. In the equatorial region of the earth the compass needle stands horizontally, on account of the equality of attraction exerted on its poles by those of the earth; but in the polar region the needle stands obliquely, that is, one end is depressed and the other end is elevated on account of the inequality of attraction exerted on its poles by those of the earth. Such a position of the needle in polar regions is technically termed the dip of the needle.

It has been found by experiments that the human body is a magnetizable object, though far inferior to iron or steel. That it is a magnetizable object is a fact that cannot be denied, for in addition to other causes there is a large percentage of iron in the blood circulating throughout all the parts of the body.

Now, as our feet are for the most part of the day kept in close contact with the surface of that huge magnet-the earth - the whole human body, therefore, becomes magnetised. Further, as our feet are magnetised by contact with the northern hemisphere of the earth, where exist all the properties of north polarity, the south polarity is induced in our feet, and north polarity, as a necessary consequence, is induced in our head. In infancy, the palms of our hands are used in walking as much as our feet, and even later on the palms generally, tend more towards the earth than towards the sky. Consequently, the south polarity is induced in them as it is at our feet. The above arrangement of poles in the human body is natural to it, and therefore conducive to our health and happiness. The body enjoys perfect health if the magnetic polarity natural to it be preserved unaltered, and it becomes subject to disease if that polarity be in the least degree altered or its intensity diminished.

Although the earth is the chief source whence the magnetism of the human body is derived, yet it is no less due to the action of oxygen. Oxygen gas being naturally a good magnetic substance, and being largely distributed within and without the human body, helps the earth a good deal in magnetising it.

Though every human, the body is placed under the same conditions with regard to its magnetisation, yet the intensity and permanence of the magnetic, polarity of one are not always equal to those of another. Those two properties of the human body are generally in direct ratio to the compactness of its structure and the amount of iron particles entering into its composition.

Now it is very easy to conceive that if you lie down with your head placed southward and feet northward, the south pole of the earth and your head, — which is the north pole of your body, and the north pole of the earth and your feet, which are the two branches of the south pole of your body, being in juxtaposition, will attract each other, and thus the polarity of the body natural to it will be preserved; while for the same reason, if you lie with your head placed northward and feet untoward, the similar poles of your body and 'the earth being in juxtaposition will repel each other, and thereby the natural polarity of your body will be destroyed or its intensity diminished. In the former position the polarity your body acquires during the day by standing, walking and sitting on the ground, is preserved' intact at night during sleep; but in the latter position, the polarity which your body acquires during the day by standing, walking and sitting on the ground is altered at night during sleep.

Now, as it has been found by experiment that the preservation of natural magnetic polarity is the cause of health, and any alteration of that polarity is the cause of disease, no one will perhaps deny the validity of the *ślokas* which instruct us to lie down with our heads placed southward, and never and nowhere to lie down with our heads placed northward.

Now, why in those two *ślokas* the eastern direction is preferred to the western for placing the head in lying down, is explained thus, "It has been established by experiments in all works on medical electricity that if a current of electricity pass from one part of the body to another, it subdues all inflammations in that part of the body, where it enters into and produces some inflammation in the part of the body whence it goes out. This is the sum and substance of the two great principles of Anelectrotonus and Catelectrotonus, as they are technically called by the authors of medical electricity.

Now, in lying down with the head placed eastward, the current of thermal electricity which is constantly passing over the surface of the earth from east to west, passes through our body also from the head to the feet and therefore subdues all inflammation present in the head, where it makes its entrance. Again, in lying down with the head placed westward, the same current of electricity passes through our body from the feet to the head, and therefore produces some kind of inflammation in the head, whence it goes out. Now, because a clear and healthy head can easily acquire knowledge, and an inflamed, or, in other words, congested head is always the laboratory of vague and distressing thoughts, the venerable sage Mārkaṇḍeya was justified in saying that man becomes learned by lying down with his head placed eastward and is troubled with distressing thoughts by lying down with his head placed westward.

There are other time honoured practices, which are founded upon a knowledge of the principles of electricity and magnetism. For instance, we find that (I) Iron or copper rods are inserted at the tops of all temples; (2) Mindulies (metallic cells) made of either gold, silver or iron, are worn on the diseased part of the body; (3) Seats made of either silk, wool, kuśa grass or hairy skins of the deer and tiger are used at the time of saying prayers. Those who are acquainted with the principles of electricity will be able to account for these practices. They know that the function of the rod of the Triśula (trifurcated iron rod) placed at the top of the Hindu temples is analogous to a lightning conductor. The mindulies perform the same functions as electrical belts and other appliances prescribed in the electrical treatment of diseases. The golden temple of Vishveshwar at Banaras is really thunder proof shelter. Professor Max Müller recommends the use of a copper envelope to a gunpowder magazine to exclude the possibility of being struck by lightning. The woollen and the skin āsanas (seats) protect our lives during a thunderstorm from the action of a return shock and keeps our body insulated from the earth.

There is another practice among the Indians. 'In representation, "around the head of each of the Indians gods is the aureole." But why they should be so represented was a mystery until long back. Baron Von Reichenbach, an Austrian chemist of eminence, thus explains it. He says, "The human system, in common with every animate and inanimate natural object, and with the whole starry heavens, is pervaded with a subtle aura, or, if you please, imponderable fluid, which resembles magnetism and electricity in certain respects, and yet is analogous with neither. This aura, while radiating in a faint mist from all parts of our bodies, is peculiarly bright about the head, and hence the aureole." "In fact," says Col. Olcott, "we see that Reichenbach was anticipated by the Aryans (Indians) in the knowledge of the odic aura". And yet "we might never have understood what the nimbus about Krishna meant, but for this Vienna chemist, so perfect is the sway of ignorance over this once glorious people."[9]

Another practice of the Indians is that "when they sit down to eat, every man is isolated from his neighbours at the feast; he sits in the centre of a square traced upon the floor, grandsire, father and son, brother and uncle, avoiding touching each other quite as scrupulously as though they were of different castes. If I should handle any individuals's brass platter, his Loṭā or another vessel for food and drink, neither he nor any of his caste would touch it, much less eat or drink from it until it had been passed through fire: if the utensil were of clay it must be broken. Why all these? That no affront is meant by avoidance of contact is shown in the careful isolation of members of the same family from each other. The explanation submitted is that every member was

---

[9] Col. Olcott's lecture delivered at the Town Hall, Calcutta, on 5th April, p. 82 quoted by Diwan Bahadur Harbilas Sarda (2007: 359)

supposed to be an individual evolution of psychic force, apart from all consideration of family relationship: if one touched the other at this particular time when the vital force was actively centred upon the process of digestion, the psychic force was liable to be drawn off, as a lead jar charged with electricity is discharged by touching it with your hand. The oldest member of old was an initiate, and his evolved psychic power was employed in the Agnihotra and other ceremonies. The case of the touching or the eating or drinking vessel, or the mat or clothing of a member by one or another caste of inferior psychic development, or the stepping of such a person upon the ground within a certain prescribed distance from the sacrificial spot, bear upon this question. Here it may also be pointed out that the aura is streaming from the points of the human hand. Every human being has such an aura, and the aura is peculiar to himself or herself as to quality and volume. Now, the aura of a person of the ancient times was purified and intensified by a peculiar course of religious training – let us say psychic training — and if it should be mixed with the aura of a less pure, less spiritualized person, its strength would of necessity be lessened; its quality adulterated. Reichenbach tells us that the odic emanation is conductible by metals, slower than electricity, but more rapidly than heat, and that pottery and other clay vessels absorb and retain it for a great while. Heat he found to enormously increase quantitatively the flow of odyle through a metal conductor. The person, then in submitting his odylically – tainted metallic vessel to the fire, is but experimentally carrying out the theory of Von Reichenbach.

The country with such a great long lasting scientific tradition and culture must have developed an advance level of science and lived it practically. Today when scholars have started thinking in terms of science in Vedas and other Sanskrit literature, the things have started changing. Modern scholars and scientists are taking interest in the scientific studies carried out in the Vedic and post-Vedic period. During their researches, they have been able to locate a vast body of scientific literature written in Sanskrit during the times of yore. The various scientific studies carried out in Vedas and allied literature and publication of scientific works in Sanskrit has made scholars more curious to search for more and more Vedic scientific literature written in the past. This search can be bi-directional. First thing is to ransack the entire collection of ancient Sanskrit manuscripts located in the libraries and museums of our country and various other countries of the world. Second thing is to register references of scientific books in the extant Sanskrit literature. The aim of the present paper is to attract the attention of the scholars interested in the study of Vedic scientific literature to collect maximum possible information in the form of manuscripts on scientific issues from the various libraries and museums of India and the other the parts of the world and information of scientific titles referred to in the extant Sanskrit texts and their commentaries.

Hereunder, we present a list of such scientific titles with or without their authors as are referred to in various Sanskrit texts and commentaries. Some of this literature has been located and documented, rest of the books have not been traced. With the help of this list scholars and readers working on ancient Indian sciences or interested in the scientific literature of the Bharat will be able to have an estimate of the numbers and types of the books written dealing with the various aspects of science and technology prevalent from ancient period till 18th century AD. They will also be obliged to trace more and more literature of this kind.

Agriculture

1. *Agatattvalaharī* -- Deals with Agriculture, methods of cultivation of plant kingdom, description of trees and their treatment. Authorship assigned to Maharishi Atri and Āśvalāyana.

2. *Kṛṣi Parāśara* - Methods of cultivation are discussed here. Authorship is assigned to Parāśara.

3. *Kṛṣi Śāstra* 4. *Kṛṣi Kaumudī*

Plants

1. *Udbhijjatattvasārāyaṇam* 2. *Udbhija Prakaraṇa*

Water

1. *Aptattatva Prakaraṇa* -- Deals with the different types of waters and their utility. The importance of bathing in different waters is mentioned. Characteristic features of different waters are described. Authorship assigned to Maharishi Āśvalāyana.

2. *Aptattva*

Cosmology/Cosmogony

1. *Aṇḍa Kaustubham* - Description of galaxies and types of living beings therein etc. is mentioned. Authorship assigned to Maharishi Parāś ara.

2. *Ākāśa Tantra* - Deals with seven types of skies, different portions of the universe, classification of stars in the sky, the interaction of various energies in the sky, types of Power, Fire, Light, the orbit of planets, Earth. Rivers and their description are also mentioned. Authorship is assigned to Maharishi Bhāradwāja.

3. *Kaumudī* - The universe is critically discussed in this work. Authorship is assigned to Somanātha.

4. *Kheṭa Sarvasva* - In this work galaxy and planetary motions have been discussed. Authorship is assigned to Maharishi Jaimīni.

5. *Brahmāṇḍa Sāra* - History of Universe is explained here. Authorship is assigned to Vyāsa.

Photography

1. *Aṁśubodhini* - Deals with the photography, planetary motions, the influence of other interfaces on their motions, light, heat sound, telephony, constructions of aeroplanes,

electricity and its applications.

Dictionary of Technical Terms

1. *Paribhā|endash ā Candrikā*

2. *Nāmārtha kalpa-sūtram* (Science of naming the different parts of the machine): In this 84,00,000 śaktis (energies) have been mentioned with their names. Also as for how they can be generated and meaning of words etc, are also described. The authorship is assigned to Maharishi Atri.

Acoustics

1. *Sarvaśabda Nibandhanam* (Acoustics)

Ayurveda

1. *Ṛk hṛdaya Tantra* - Different types of diseases, their treatments are described therein. Authorship assigned to Maharishi Atri.

2. *Ayurveda Prakāśa*

Arthaveda

Arthaveda

Animal Sciences

1. *Aśva Śāstra* 2. *Gaja Śāstra* 3. *Aśva Lakṣaṇa Sāra*

4. *Mṛga cāramiya* 5. *Hayalīlāvati* 6. *Go Śāstra*

7. *Kukkuṭa Śāstra* 8. *Aśva Tantra* 9. *Gaja Tantra*

10. *Gorakṣā Tantra*

Dhanurveda

1. *Dhanurveda* by Maharishi Vasiṣṭha

2. *Astra Vidyā* 3. *Vyuha Śāstra* 4. *Senā Śāstra*

5. *Rathaśīkṣāśāstra* 6. *Sūta Śāstra*

7. *Vāhanarohaṇa Śāstra*

8. *Kamandaka* 9. *Dhanu Śāstra* 10. *Bhojarājīva*

11. *Revantottara* 12. *Vyuhua Lakṣaṇa* 13. *Senā Lakṣaṇa*

Marshal Arts

1. *Malla Śāstra*
2. *Malla Vidyā Prakāśa*
3. *Weapons/Armoury*
4. *Nālikānirṇaya*

Space

1. *Kheṭa-vilasa-grantha* (Flying in Sky)
2. *Ākāśatantram of Bharadwāja*

Fire

1. *Vaiśvānara Tantra* - 128 types of fire, their colours, behaviour, uses, measurements, mutual differences etc. are described. Authorship is assigned to Maharishi Nārada.

Mechanics/Machinery

1. *Śuddha Vidyākalpa* - This describes different types of machines including for making an aeroplane and sound machines. Authorship is assigned to Āśvalāyana.
2. *Bṛhadyantra sarvasva* - Explains all types of machines including that of aeroplanes and electricity. Authorship is assigned to Maharishi Bharadwāja.
3. *Yantra Kalpa* (Flying Machines) by *Maharishi Garga*
4. *Yantra Saṅgraha*
5. *Yantra Kalpataru* (Mechanical Engineering)

Aeronautics

1. *Samarāṅgaṇa Sūtradhāra* - This describes the manufacturing of airplanes with mercury as their fuel. Authorship is assigned to King Bhoja of Malwa.
2. *Vimāna Candrikā* by Nārāyaṇa
3. *Vyomayāna Tantra* by Maharishi Śaunaka
4. *Yāna bindu* by Vacaspati
5. *Kheṭayāna Pradīpikā* by Cakrāyaṇa
6. *Vyoma-yāna-arka-prakāśa* by Ḍhuṇḍīnātha
7. *Vyoma-yānārka Sikṣā*
8. *Kriyāsāra* (Practical exercises of the Aeronautics)
9. *Śaunakīyam* (Work of Śaunaka on Aeroplanes)

10. *Kheṭa sarvasva* 11. *Kheṭa Yantram*

Instrumentation

1. *Yantraprakaraṇa* (Installation of instruments in planes)
2. *Maṇiratnākara*
3. *Maṇibhadra Kārikā* (Installation of Maṇis in Planes)
4. *Maṇi Prakaraṇa* (Installation of Maṇis in Planes)
5. *Maṇi Kalpa Pradīpikā*

Air spring

1. *Cāra-Nibandana grantha*

Air conditioning in Aeroplanes

1. *Ṛtukalpa*

Colour

1. *Varṇsarvasvam* (Colour-management)

Meteorology

1. *Karaka Prakaraṇa* - Deals with changes in clouds, changes in sun-rays, and the relationship of clouds. It also discusses the role of sun-rays in generating precious stones. (Navaratnas). Authorship is assigned to Aṅgirasa.
2. *Meghotpatti Prakarṇa* - Types of clouds, thunder, lightning, and their effects are explained here. The changes in clouds, generation of the life of many species, changes in solar energy, the relationship between solar radiation and the cloud formation, the origin of Navaratnas and how the solar radiation is responsible for their origin. Authorship is assigned to Maharishi Aṅgirassa.
3. *Saudāminī Kalā* (Science of lightning)

Metallurgy

1. *Dhātu Sarvasva* - In this work elements, minerals, metals, alloys and their extraction from mines by different methods are discussed. This also deals with poisons and antidotes, production of mercury (since mercury was also used as a fuel in machinery), sulphur etc and preparations of ashes. Authorship is

assigned to Maharishi Baudhāyana.

2. *Lauha Tantra* - Ores, their genesis and extraction are discussed here. Authorship is assigned to Śākaṭāyana.

3. *Lauha Tattva Prakaraṇa*4. *Lauha Ratnākara*

5. *Lauha Rahasyam* 6. *Lauha Śāstram* of Śākaṭāyana

7. *Lauha Paddhati*

8. *Lohasarsvam* (Work on metallurgy of planes)

9. *Dhātusarvasvam* (Work on aeronautical metallurgy)

11. *Lohaprakaraṇa* (Work on metallurgy of planes)

12. *Lauhādhikaraṇam*

Science of Smoke/Vapour

1. *Dhūma Prakaraṇa* - This work discusses in detail different types of smokes/vapours. The detection of vapours/smokes with the help of mirrors. Research on various types of vapours/smokes with acids to find out whether the vapour is harmful or not for mental and physical growth. Authorship is assigned to Nārada.

2. *Āpa Tattva* - In this 84,000 vapours, their layers, their impact on earth and plantation, 84,00,000 medicines and instruments to detect these vapours are described. The authorship is assigned to Maharṣi Śākṭāyana.

Creation

1. *Prapañca Laharī* - This work discusses atom in detail. The question whether this Universe is created by atoms or by Brahma-tattva is addressed here. Authorship is assigned to Vasiṣṭha.

2. *Sṛṣṭi Vilāsa*

3. *Prapañcasāra*

Language

Loka Saṅgraha - This work discusses 1714 languages, living beings, their origin, and food habits and different information of the world. Authorship is assigned to Vivarṇācārya.

Air

Yāyu Tattva Prakaraṇa - This work discusses ionosphere,

different layers in it. 4000 varieties of gases and effects of these gases on the earth, on flora and fauna. The instruments to detect them in the atmosphere are explained in detail. Authorship is assigned to Maharishi Śākaṭāyana.

Chemistry

1. *Rasa-ratna-samuccaya* - This work discusses chemistry. Different chemicals like mixed chemicals, compound chemicals, ordinary chemicals, their actions and reactions with different metals have been discussed. Different types of machinery, metallurgy, elements and their characters have been explained. Metals like Gold and Silver etc. their characters and their different kind of treatments to the human body have been discussed in detail. Authorship assigned to Vāgabhaṭṭācārya.

2. *Rasa Ratnākara*- This book is divided into five *Khaṇḍas:* 1. *Rasa Khaṇḍa* 2. *Rasāyaṇa Khaṇḍa* 3. *Rasendra Khaṇḍa* 4. *Vāda Khaṇḍa* 5. *Mantra Khaṇḍa.* The authorship is assigned to *Nityanātha Siddha.*

3. *Rasendra Cuḍāmaṇī* by Somadeva. 4. *Rasendra Cintāmaṇi*

5. *Rasa Saṅketa Kārikā\tab*

6. *Rasendra Maṅgalam by Nagarhjuna*

7. *Rasa Sāra* 8. *Rasanāmadhenu*

9. *Rasa Kaumudi* 10. *Rasendra Vijñāna*

11. *Rasendra Sāra Saṁgraha* 12. *Rasādhyāya*

13. *Rasopaniṣat* 14. *Ānanda-kāṇḍa*

15. *Rasa-paddhati* 16. *Rasa Kāmadhenu*

Power/Electricity

1. *Śakti Tantra* - This describes electricity and its powers such as *sarvākarṣa, rūpākarṣa, rasākarṣa, gandhākarṣa, śabdākarṣa, dhairyākarṣa, śarīrākarṣa, prāṇākarṣa,* and others (a total of 16 types of powers) are explained. Authorship is assigned to Maharishi Agastya.

2. *Nāmārdha Kalpa* - This work defines 8.4 Million types of powers and gives their nomenclatures. Authorship is assigned to Maharishi Atri.

3. *Śakti Bījam* (Power Supply in Machines)
4. *Śakti Kaustubha* (Power Management)
5. *Śakti Sarvasva* (Power management in Planes)
6. *Śakti Vilāsa*
7. *Śakti tantram* (Work on Power management in planes)
8. *Śaktisūtram of Agastya*
9. *Rūpaśakti Prakaraṇam of Aṅgiras*

Speed Management in Machines

1. *Saṁskāra Ratnākara* (Speed management in Aeroplanes)
2. *Gatinirṇayādhikāra*
3. *Drāvaka Prakaraṇa*
4. *Gatinirṇayādhyāya* (Speed management)

Sound Management in Machines

1. *Śabda Mahodadhi* (Sound management in Aeroplanes)
2. *Śabda Nibandhanam* (Sound control in machines)

Science of Rays

1. *Āśani Kalpa* 2. *Aṁśum Tantram* of Bharadwāja
3. *Aṁśubodhinī*

Glass Technology

1. *Darpaṇa Kalpa*
2. *Darpaṇa Śāstram*
3. *Darpaṇa Prakaraṇa* (Mirror Planning in Aeroplanes)
4. *Mukura Kalpa* (Mirror Technology)
5. *Darpaṇa Saṅgraha* (Mirrology)

Witchcraft

Sammoha Kriyākāṇḍam (How to make an enemy unconscious)

Gemology

1. *Suvarṇa Ratnādi Parikṣā|tab* 2. *Navaratna Lakṣaṇa*

Commerce

Vāṇijya Śāstra

Geology

Bhū Parikṣā Śāstra

Ethics

1. *Nīti Śāstra*
2. *Nīti Sāra*
3. *Nīti Sūtra*
4. *Nīti Kalpataru*
5. *Nīti Prakāśikā*
6. *Nīti Mañjarī*
7. *Nīti Candrikā*

Military Science

Samara Sāra

Pharmacy

Auṣadhi Kalpa of Atri

Artha Śāstra

1. *Artha Śāstra* by Kauṭilya
2. *Artha Śāstra* by Bṛhaspati
3. *Artha Śāstra* by Bharatṛhari
4. *Rājanīti Ratnākara*

Political Science

1. *Nāgara Sarvasva*
2. *Daṇḍa Nīti Śāstra*
3. *Dhaumaya Nīti Śāstra*
4. *Akṣnīti Sudhākara Gaṇita* (Mathematics)

Mathematical Sciences

1. *Aṅkagaṇita*
2. *Bījagaṇita*
3. *Kṣetragaṇita*
4. *Chāyāgaṇita*
5. *Kūpādigaṇita*
6. *Bhinnādhikāra*
7. *Prakīrṇagaṇita*
8. *Trairasikagaṇita*
9. *Miṣragaṇita*
10. *Ghaṭagaṇita*
11. *Vāstugaṇita*
12. *Sūtragaṇita*

13. *Suvarṇagaṇita*

14. *Vālamīki gaṇita*

Architecture

1. *Brahmāṇḍa Vāstu*

2. *Rājadhāni Nagar Nirmāṇa vāstu*

3. *Pattana Vāstu* 4. *Grāma vāstu*

5. *Nāvika Vāstu* 6. *Sthāpatya Tantra*

Geography

1. *Bhūgola* 2. *Dvīpa Viveka*

Toxicology

Viṣanirṇayādhikāra (Air -Texicology)

Food Technology

Aśana Kalpa (Food Technology)

Tempering Technology

1. *Pāka-sarvasva* 2. *Pārthiva-pāka-kalpa*

3. *Saṁskāra Darpaṇa* (Tempering and heating technology)

Thermodynamics

Bhastrikānibandhanam

Textile Engineering

1. *Paṭakalpa* 2. *Kṣirīpaṭakalpa*

3. *Paṭapradīpikā\tab* 4. *Paṭṭasaṁskāra Ratnākara*

(Textile Technology)

Miscellaneous Books

1. *Drāvaka Prakaraṇa*

2. *Śucivaṇa Karma* 3. *Nidhi Pradīpa*

4. *Abhilaṣitārtha Cintāmaṇī* 5. *Yukti kalpataru*

6. *Paribhā\endash ā Candrikā\tab*

7. *Nāmartu Kalpa* by Atri

8. *Dhvānta Vijñāna Bhāskara*
9. *Akṣa Tantra* by Atri
10. *Nāgārjuna Tantra* by Aiśvara
11. *Māṇḍvya Tantra*
12. *Vyāḍi Tantra*
13. *Patañjali Tantra*
14. *Kapiñjala Tantra*
15. *Bhṛgu Tantra*
16. *Agastya Tantra*
17. *Aprājita Pṛcchā|tab*
18. *Aneka Vidyākalpa Nirupaṇādhyāya*
19. *Arya Vidyā Sudhākara*
20. *Mānasāra*
21. *Mānasollāsa*
22. *Sāmarājya Lakṣmī Pī|emdash aka*
23. *Avyāja Palini Tantra*
24. *Rajyalakṣamī prasādaka Tantra*
25. *Aṅgarāgamana Tantra* (How to walk on the burning coals?)
26. *Rājyopāṅga tantra*
27. *Stribalarāja Tantra*
28. *Cāmaradvandva Tantra*
29. *Praharṣarāja Tantra*
30. *Dhīra Tantra*
31. *Grāmapālana Tantra*
32. *Kāmarūpa Tantra*
33. *Kālajivanika Tantra*
34. *Kuñja Kambala Tantra*
35. *Kanakaprayoga Tantra*
36. *Bodhini Tantra*
37. *Pādukā Tantra*
38. *Kheṭaka Tantra*
39. *Lepa Tantra*
40. *Śalya Tantra*
41. *Pātāla Gamana Tantra*
42. *Cakrāyuddha Prayoga Tantra*
43. *Lekhā Tantra*
44. *Pṛthivi Tantra*
45. *Śulka Tantra*
46. *Giri Tantra*
47. *Vana Tantra*

48. *Bodhānanda Kārikā* of Bodhānanda

49. *Viśvambhara Kārikā of Viśvambhara*

50. *Mūlārkaprakāśikā*\tab 51. *Śaṇaniryāsa Candrikā*

52. *Bṛhatkāṇḍam* 53. *Paṭikānibandhana*

54. *Nirṇyādhikāra* (Decision making)

55. *Mū*\endash a *Kalpa* 56. *Kuṇḍakalpa*

57. *Kuṇḍanirṇaya* 58. *Parāṅkuśa*

59. *Niryāsa-kalpa* 60. *Pralaya Paṭalam*

61. *Ṣaḍagarbha viveka* 62. *Raghūdaya*

63. *Jīva sarvasvam* of Jaimini

64. *Karmābdhipāra* of Āpastamba

65. *Chandaḥ Kaustubha* of Parāśara

66. *Kaumudī* of Siṁhakhoṭha

67. *Ṛkhṛdayam* of Atri

68. *Lokasaṅgraha* of Visaraṇa

Apart from the above-cited treasure of Vedic scientific literature, we also come across the names of some 36 eminent Vedic scientists who contributed brilliantly in various scientific fields during the Vedic period. They may be enumerated as under:

1. Nārāyaṇa Muni	2. Śaunaka	3. Garga
4. Vācaspati	5. Chākrāyaṇi	6. Dhuṇḍinātha
7. Viśvanātha	8. Gautama	9. Lalla
10. Viśvambhara	11. Agstya	12. Buḍila
13. Gobhila	14. Śākaṭāyana	15. Atri
14. Kapardī	17. Gālava	18. Agnimitra
15. Vātāpa	20. Sémba	21. Bodhānanda
22. Bhardwāja	23. Sidhnātha	24. Īśvara
25. Āśvalāyana	26. Vyāsa	27. Parāśara
26. Siṁhakoṭa	29. Aṅgirā\tab	30. Visaraṇa

27. Vasiṣtha	32. Jaimini	33. Āpastamba
34. Baudhāyana	35. Nārada	36. Vālmīki

References

Albrecht Weber, 1878. *History of Indian Literature*, Trübner & Company, London

Cuvier, 1835. The *Theatre of the Hindus*, Vol. I, Parbury, Allen and Company. London

Diwan Bahadur Harbilas Sarda, 2007. *Hindu Superiority*, Ed. Dr. Ravi Prakash Arya, International Vedic Vision, New York & Indian Foundation for Vedic Science, India.

Elphinstone M. 1843. *History of India*, Vol. 1, J. Murray London

George C. Swayne, 1887. *Herodotus*, published by William Blackwood and Sons, Edinburgh and London

Manning (Mrs.), 1869. *Ancient and Mediaeval India*, Vol. II, W.H. Allen & Co. London.

Praphula ChandraRay. 1902. *History of Hindu Chemistry*, Vol. I. Williams and Norgate, London

www.ingramcontent.com/pod-product-compliance
Lightning Source LLC
Chambersburg PA
CBHW060514200326
41520CB00017B/5038
9788187710899